98

j635
CRE
　　　　Creasy, Rosalind.
　　　　　　Blue potatoes,
　　　　orange tomatoes

Blue Potatoes, Orange Tomatoes

Blue Potatoes, Orange Tomatoes

by Rosalind Creasy

Illustrations by Ruth Heller

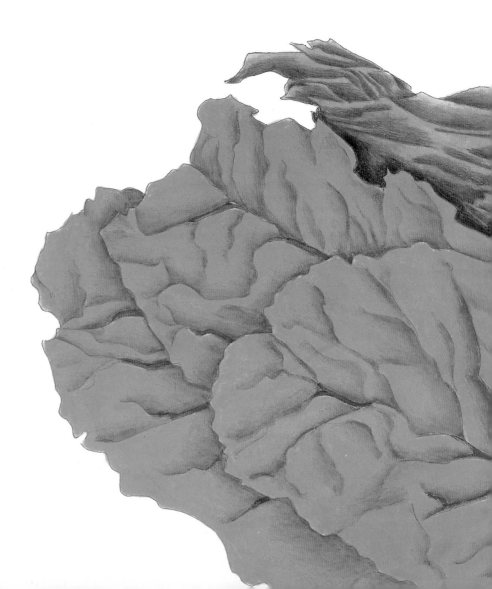

Sierra Club Books for Children
San Francisco

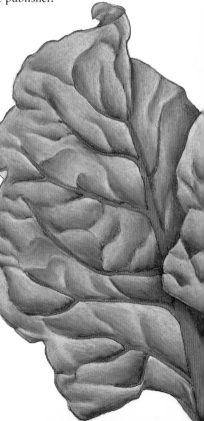

For Noah and Sierra   R.C.

To Julia   R.H.

The Sierra Club, founded in 1892 by John Muir, has devoted itself to the study and protection of the earth's scenic and ecological resources — mountains, wetlands, woodlands, wild shores and rivers, deserts and plains. The publishing program of the Sierra Club offers books to the public as a nonprofit educational service in the hope that they may enlarge the public's understanding of the Club's basic concerns. The Sierra Club has some sixty chapters in the United States and in Canada. For information about how you may participate in its programs to preserve wilderness and the quality of life, please address inquiries to Sierra Club, 730 Polk Street, San Francisco, CA 94109.

FIRST EDITION

LIBRARY OF CONGRESS CATALOGING-IN-PUBLICATION DATA

Creasy, Rosalind.
    Blue potatoes, orange tomatoes : how to grow a rainbow garden / by Rosalind Creasy ; illustrations by Ruth Heller. — 1st ed.
        p.   cm.
    Summary: Describes how to plant and grow a variety of colorful vegetables, including red corn, yellow watermelons, and multicolored radishes.
    ISBN 0-87156-576-5
    1. Vegetable gardening — Juvenile literature. 2. Vegetables — Varieties — Juvenile literature. 3. Cookery (Vegetables) — Juvenile literature. [1. Vegetable gardening. 2. Vegetables. 3. Gardening.] I. Heller, Ruth, 1924- ill. II. Title.
SB324.C68   1994
635—dc20   92-38800

Book and jacket design by Paula Schlosser

Composition by Harrington-Young Typography

Printed in Hong Kong

10   9   8   7   6   5   4   3   2   1

Contents

Growing Your Rainbow Garden

Everybody knows that all string beans are green and all tomatoes are red — right?

Not so! *Usually* string beans are green. But they can also be purple, maroon, or yellow. Tomatoes can be orange, yellow, or even purple. Lots of familiar fruits and vegetables grow in a whole rainbow of surprising colors.

A rainbow garden is great fun to grow. I always smile when I fill my basket with red popping corn, or pull multicolored radishes out of the ground. What color will the radishes be? I never know until I harvest them. And I love to amaze my friends and family with my blue potatoes. They think I've dipped them in food coloring!

If *you'd* like to grow vegetables and fruits that are just a little bit different, why not try planting a rainbow garden of your own? Rainbow crops are easy to grow, and you don't need a huge plot of land. In a space about the size of the bed you sleep in, you can plant two orange tomato plants, two yellow zucchini plants, and a short row of purple string beans, with multicolored radishes in between. If you plant some golden marigolds, purple pansies, and orange nasturtiums around the edges, your garden will be even prettier to look at.

In the pages that follow, you'll find everything you need to know about growing a rainbow garden. The first part of the book gives you all the basics: how to order seeds, how to get your garden ready, and how to plant and care for your growing plants. The second section offers a rainbow of individual fruits and vegetables to choose from.

And when it's time to harvest your yellow watermelons, blue potatoes, and lavender radishes, I hope you'll have as much fun serving them to family and friends as I do. My red chard looks beautiful on a salad plate. And when I make red, white, and blue potato salad for the Fourth of July, everyone cheers. Who knows — once you've tried the recipes in this book, maybe you'll be inspired to invent your own.

To get started on your rainbow adventure, just turn the page. Happy planting!

First, the Seeds

When you're a rainbow gardener, the seeds for your garden won't be just ordinary ones. They'll be special varieties you might not find at your local nursery. That means you'll need to get them from mail-order seed companies like those listed on the next page.

For a head start on your garden, begin looking at seed catalogs during the winter. That way you'll be ready to order your seeds by early spring.

To get the catalogs, just write the company a short letter. An example is shown on the opposite page. If the catalog costs a dollar or two, ask an adult to write a check; don't send cash.

Your catalogs will probably arrive in two to three weeks. They'll be chock-full of illustrations and exciting ideas. Some may even have projects and recipes. You'll have fun just browsing through them. Each type of crop — potatoes, tomatoes, beans, and so forth — comes in many different varieties. And each variety has a name, like "All-Blue" potatoes, or "Royalty" string beans. To get an idea of the varieties you may want to consider, look at the second section of this book.

Carefully list the varieties you want on the order form that comes with the catalog. Ask an adult to write a check for the total amount, and mail it off with the order form. You can expect to get your seeds in two to three weeks.

January 10, 1999

Seeds Blum
Idaho City Stage
Boise ID 83706

Dear Seeds Blum

Please send me a copy
of your seed catalog
so that I can order
seeds for my garden.
Enclosed is my check
for $2.00.

Thank you

Chris Smith
1111 A Street
Anyplace, CA. 00000

Seed Companies

W. Atlee Burpee Company
Warminster, PA 18991

Harris Seeds
60 Saginaw Drive
Rochester, NY 14623

Johnny's Selected Seeds
Foss Hill Road
Albion, ME 04910

Nichols Garden Nursery
1190 North Pacific Highway
Albany, OR 97321

Park Seed Company
Cokesbury Road
Highway 254 North
Greenwood, SC 29647

Seeds Blum
Idaho City Stage
Boise, ID 83706
(Catalog: $3.00)

Shepherd's Garden Seeds
30 Irene Street
Torrington, CT 06790
(Catalog: $1.00)

Stokes Seeds, Inc.
P.O. Box 548
Buffalo, NY 14240

9

A Home for Your Plants

Your first job as a rainbow gardener is to provide a good home for your plants. After all, if your tomatoes get cold, they can't wrap up in a blanket. And if the soil gets too soggy, your zucchini plants can't pull up and tiptoe around looking for a drier place. *You* have to put them in the right spot to begin with.

For a plant, a comfortable home has:

- the right temperature
- lots of sunshine
- a soft, fluffy bed of soil

Some like it hot, some like it cool

Each vegetable or fruit grows best in a certain kind of *climate* — a combination of temperature, moisture, and sunlight. Some plants love hot temperatures day after day; these are the warm-season plants. Others need lower temperatures; these are the cool-season plants. Clever gardeners have figured out a simple rule for remembering which is which. To see how it works, look at the charts below (this page and opposite).

With both warm- and cool-season plants, you need to be aware of when frost comes to your area. Frost is a covering of tiny ice crystals that form on plants in cold weather. Tomato,

If the fruit* or seed is what you eat, it's usually a
Warm-Season Plant
It likes: 70°–80°F (21°–27°C); 80°–100°F (27°–38°C) for watermelon
It grows in: Summer
Examples in this book: Tomatoes; corn; string beans; zucchini; watermelon
*The fruit is the part of the plant that holds the seeds.

corn, string bean, zucchini, and potato plants will turn brown and die if they get any frost on their leaves. But radishes and chard can stand some frost. Ask the people at your local nursery to tell you the date of your area's average last frost in spring and average first frost in the fall. Then you'll be sure not to plant too soon in the spring, or harvest too late in the fall.

Sunny days

The next important thing your fruits and vegetables need is lots of sunshine. All the crops in this book need at least six hours of bright sun every day to grow well. So be sure to plant them in a spot in your garden that gets plenty of sun.

If the root, tuber,* or leaf is what you eat, it's usually a

Cool-Season Plant

It likes: 50°–70°F (10°–21°C)
It grows in: Spring, fall, winter (except where it gets icy cold)
Examples in this book: Radishes (roots); potatoes (tubers); Swiss chard (leaves)
*A tuber is an underground stem.

A nice bed of soil

Every fruit and vegetable plant needs a bed of rich, fluffy soil to sink its tender little roots into. The roots need to breathe and take in water and minerals from the soil so the rest of the plant can grow. That means you can't just dig a hole in your yard and plant your seeds. You have to get a special plot of soil ready first. It should be loose and soft, with plenty of air pockets for the roots to wiggle into. And it has to be full of food, called *nutrients*, in the form of *compost* or animal manure. (Compost is a material made from decomposed garden clippings, leaves, and even kitchen scraps.)

How big does your garden plot have to be? It all depends on how much space you have and what you want to grow. For a handful of radish plants and some string beans, you need a garden bed only 8 to 10 feet long by 2 feet wide. Even just a few large barrels or wood planter boxes will do.

To grow a bigger garden — say, a few corn plants, two zucchini plants, and three or four tomato plants — you need a bed measuring about 10 feet by 10 feet. The pages about the individual plants in the second part of this book will tell you how much space they should have.

Rows or hills?

Some crops, such as potatoes, string beans, tomatoes, and radishes, grow well in rows. Others, including watermelon, zucchini, and corn, need little hills — round piles of soil about 3 feet wide and 1 foot high. You may want to make a combination of rows and hills for your garden. Before you begin, round up all the necessary tools — then start digging!

To make *rows,* dig up an area the total size you'll need. (If the ground is too hard, water it well and then come back and dig it the next day.) Dig down 6 to 8 inches with your shovel. Pull out any weeds, and use a spading fork to break chunks of soil into little pieces.

Spread a layer of manure or compost 3 to 4 inches thick over the whole bed. Use your shovel or garden fork to mix it into the soil. Finally, smooth the bed with a rake. To make the rows, you can simply trace them into the dirt with your fingers.

To make a *hill*, first dig a hole 3 feet across and 6 to 8 inches deep. Pile the soil next to the hole and then break up any chunks and take out weeds. Mix compost or manure (about enough to fill two grocery bags) into the soil pile. Then shovel the soil back into the hole. Since it won't all fit, the extra becomes the hill.

The plants that grow in hills tend to get large, so if you make more than one hill, be sure to place them at least 4 feet apart.

Tools You'll Need

Garden shovel
for digging

Spading fork for
breaking up soil

Rake for
smoothing soil

Trowel for planting

Watering can

Now, for Planting

You have a whole collection of little paper packets full of rainbow fruit and vegetable seeds. Now it's time to get them started. You may be able to start them right out in the garden, or you may have to start them indoors — it depends on which crops you're growing and what your climate is like.

Indoor starters

Some plants take a long time to grow, but they can't be started outside too early in the year because of the danger of frost. Tomatoes are best started indoors. If you live in a short-summer area, you also have to start corn, watermelon, and zucchini indoors. Check the individual fruit and vegetable pages in the second part of this book to figure out which plants you'll need to start this way.

Let's say you're going to start two tomato plants. You'll need:

- two clean 1/2-gallon milk cartons or two 6-inch flower pots

- a bag of potting soil from the nursery

- an old tray to put under your containers to catch the drips

- a warm, sunny spot inside your house

If you're using milk cartons, cut the tops off. Make a few small drainage holes in the bottom of each carton. Then fill your cartons almost to the top with soil.

Place six tomato seeds an inch or so apart in each container. Gently push the seeds into the soil until they are down as deep as your first knuckle. Smooth the soil back over the seeds and pat it down gently. Sprinkle each container with 2 cups of water and let drain in the sink.

Label your plants by writing the plant name on a Popsicle stick with a ballpoint pen and sticking the label into the dirt at the *edge* of the container.

Put the containers in their tray by a warm, sunny window. The seeds will sprout fastest if you keep them between 75° and 85°F (24° and 29°C).

Sprinkle the containers with water every few days. The soil should be damp, not soaking wet.

Your seeds should start sprouting in about ten days. Once the little plants come up, make sure they're in a place that gets sunshine all day. When the seedlings are about 2 inches tall, carefully remove all but one healthy plant in each container. (It sounds mean, but it will help the *one* plant grow best.) Fertilize with fish emulsion (see page 18).

Out into the garden

Your tomato plants (or any other plants you start indoors) are ready to go outside when they're at least 3 inches tall, and at least four weeks after your average last frost date. A week or so before you're going to move, or *transplant*, them, start putting the containers outside for a few hours of morning sunshine each day. This helps the plants get used to the outdoors.

When you're ready to plant, prepare the soil well. Check the individual plant pages to see whether to plant in rows or hills. For each plant, make a hole in the soil a little bigger than the roots. As you take your plants out of their containers, try to keep the tender little roots surrounded by soil. Be careful not to break or crush them.

Put each plant in its hole, cover its roots with more soil, and pat down firmly. Then gently water each plant well.

If it doesn't rain, water your plants once a day for three or four days. Skip a day or two and water again. After that, the plants need watering (by you or by the rain) once a week for the rest of the growing season.

Outdoor starters

Plants such as radishes and string beans grow fast enough that you can wait until after your average last frost date and start the seeds right out in the garden. First, prepare the soil well. Turn to the page that tells about the kind of seeds you're planting. There you'll find out how far apart and how deep to plant them, and whether they go in rows or hills.

Just place the seeds in the dirt; then cover them with a little soil and pat them in place. Water them gently with a watering can. To keep track of what you've planted, mark the rows or hills with labels from your local nursery or ones you make yourself from used Popsicle sticks.

If it doesn't rain, water the seeds every day. One day you'll see little green sprouts poking out of the ground. After that, water your plants every few days until they are 2 or 3 inches tall. Then make sure they get watered by you or by the rain once a week for the rest of the growing season.

It's exciting to see all the seeds you've planted sprout up out of the soil! But if you let *all* of them keep on growing, they'll soon have to fight one another for water, food, and space. So you'll need to *thin* them by pulling out some of the tiny plants. You might not like doing it, but the plants that are left will be much healthier. To learn how and when to thin, see the individual fruit and vegetable pages in the back section of this book.

Tender Loving Care

Now that your plants are growing, you're on your way to having a successful rainbow garden. But of course you can't just sit back and watch it grow. A garden needs tending. All young plants need:

- feeding
- watering
- weeding
- protection from pesky insects, birds, and animals

Feeding

Vegetable and fruit plants usually can't get all of their food from the soil, so you'll have to feed them with fertilizer. I like to use fish emulsion fertilizer, which you can buy at a nursery. It smells awful, but the plants like it. To use it, run a gallon of water into a large watering can or bucket. Add 3 tablespoons of fish emulsion and stir, until it's well mixed in. Water each plant well with the fertilizer mixture. The individual plant pages will tell you how often to feed your plants.

Watering

The rainbow crops in this book need enough water to keep their roots wet. Sometimes the rain takes care of this, but if it doesn't rain enough, you must be sure your plants get plenty of water once a week. I like to use a watering can for this job, because it waters gently, without washing away the soil. A garden hose with a spray nozzle also works well.

Weeding

There's one problem that almost all gardeners have in common: weeds. Weeds are strong; they can get so pushy that they will crowd your crops until your tender little plants die. So you should start pulling out weeds when your plants are tiny. At first, you may need an adult to help you tell the difference between the new little plants and the weeds — they can look surprisingly alike!

Weed again when your plants are about 6 inches tall. After this weeding, it's time to spread *mulch* all around your plants. Mulch can be leaves, straw, or compost. You'll find many kinds of mulch for sale at your local nursery. Spread a layer 3 to 4 inches deep, and it will keep new weeds down, as well as keeping the soil soft and cool.

• FISH •
EMULSION

Critter Control

I love my garden because it's like a miniature zoo. Bees, butterflies, and humming-birds sip nectar from my flowers. Robins look for worms, and spiders weave webs that sparkle with dew. Most of the time my mini-zoo is peaceful, and I don't mind sharing some of my plants with the insects and other animals.

But sometimes garden visitors get greedy. When the bluejay starts gobbling up my new corn or radish plants, I say, "Knock it off!" I cover the seeds or plants with bird netting so the jay can't reach them. Or when large tomato hornworms eat too many of my tomatoes, I pick them off with my fingers and throw them away. (They won't hurt you; they just wiggle a lot.)

20

I don't like to spray poisons on food I'm going to eat, so I'm very careful about what sprays I use to control greedy insects. And you should be, too. If you can't pick off the insects, buy some soap spray at a nursery. This kind of spray won't hurt you or the plants, and usually it will give you all the help you need. If not, look for other solutions in books on organic gardening.

Sometimes larger critters such as rabbits, gophers, and deer will make your yard their home. They have big appetites and must be kept out. If they're a problem, ask an adult to help you, since it may require a project such as building a fence or a raised planting bed.

21

A Rainbow of Fruits and Vegetables

On the following pages, you'll find a whole rainbow of crops you can grow, from blue potatoes to yellow watermelon. For each vegetable or fruit, you'll learn what variety to order, when and how to plant, what garden critters to watch out for, and when to harvest. No matter which ones you decide to plant, remember to follow these simple rules:

- Before you plant, always read the general directions for soil preparation and planting given in the first part of this book.

- When you plant or transplant in spring, wait until the weather has been warm and sunny for a few weeks and all danger of frost is past.

- Once your plants have gotten a good start, be sure to water them weekly unless it rains a lot in your area.

When harvest time comes and you want to serve the results of all your hard work, you can use the recipes included in this section for each rainbow fruit and vegetable. You'll find tasty muffins made with golden zucchini; a creamy dip that makes stars of your multicolored radishes; and a fresh fruit salad that shines with sunny yellow watermelon balls — to name just a few.

You'll notice, too, that the recipes don't call for a lot of fancy ingredients. When it comes to fixing fresh garden produce, I've found that *simple* is usually best. Fruits and vegetables that aren't cooked too much or loaded with butter or salt keep their natural good flavors and nutrition.

All the recipes included here are easy and quick to prepare. If you don't have much cooking experience, though, be sure to ask an adult to help you with the steps that involve using a sharp knife or the stove.

Just think how proud you'll be when you carry your special dishes to the table to serve your family and friends. There's nothing quite like eating fresh, wholesome food that you've grown and tended in your own garden and prepared all by yourself.

Blue Potatoes

Most people eat plain old white potatoes. But why settle for those when you can have lavender mashed potatoes? Or, for the Fourth of July, red, white, and blue potato salad?

To get "All-Blue" potatoes, write to Seeds Blum. When your package arrives, you'll find slightly wrinkled potatoes inside instead of seeds. (Although most fruits and vegetables start from seeds, potatoes start from potatoes.) Some of your starter potatoes may be in pieces; others will be whole. They may have white or purple sprouts, or dimples called *eyes*. When you plant them, the eyes will become sprouts, and the sprouts will become plants.

Keep the package in a cool, dark place until it's time to plant. To grow sixty to a hundred potatoes, dig two rows 2½ feet apart. Each row should be about 10 feet long, 2 feet wide, and 4 inches deep.

If your starter potatoes are whole, ask an adult to help you cut each one into two or three chunks. Be sure each chunk has at least two sprouts or two eyes. Plant the chunks just as you would seeds, about a foot apart along your rows. Water the rows well after planting. If you don't get rain, water once a week until the plants turn brown at the end of the growing season.

When the plants are about 6 inches tall, fertilize them and sprinkle compost 2 inches deep on the ground around each plant. The compost will cover any little potatoes that try to poke their way out of the ground. (Uncovered potatoes will sunburn and turn green — and green potatoes are poisonous to eat!) Fertilize every four weeks until the leaves start to turn brown.

24

POTATO BUG

Sometimes you'll find potato bugs (see the picture on the opposite page) munching on your plants. If you do, just pick them off, seal them in a plastic bag, and throw the bag away.

If you want some little "new" potatoes, you can begin to harvest about fourteen weeks after planting, when the plants have flowered and the flowers have fallen off. Poke your fingers into the soil under the plants and wiggle them around to find your treasure of 2-inch potatoes. Carefully dig up a few. Then, if you keep the plants watered and fertilized, you can harvest as often as you like for about two or three weeks, as the potatoes grow larger and larger.

When the plants start to turn brown, dig up the rest of the potatoes. (It's best to use a spading fork instead of a shovel.) Take the potatoes inside and store them in a cool, dark place. They will keep for a couple of months.

Red, White, and Blue Potato Salad

2 large or 4 medium-size blue potatoes, scrubbed
2 large or 4 medium-size white potatoes, scrubbed
½ red bell pepper, washed

½ cup Homemade Salad Dressing (see page 33) or bottled Italian salad dressing

Fill a large pan with water. Add all the potatoes and bring to a boil; boil until they are almost soft when poked with a fork. Drain them and let cool slightly.

Cut the potatoes into ¼-inch-thick slices and put them into a medium-size bowl. Cut the red bell pepper into small pieces and add to the potatoes. Pour the salad dressing over the vegetables and gently toss with a wooden spoon. Refrigerate until cold (at least an hour). Serves 4.

Orange Tomatoes

Bright red tomatoes are most people's favorites. But orange tomatoes are just as tasty and colorful, and a little bit different. To grow orange tomatoes, send for "Golden Jubilee" (available from Burpee and Seeds Blum) or "Mandarin Yellow" (available from Shepherd's).

Two plants of orange tomatoes are usually enough for a family of four. Because the plants need a long time to grow juicy tomatoes, start them inside. It's best if the plants are at least 3 inches tall when you transplant them into the garden (see page 16). Place them at least 2 feet apart in a short row, and water them well. After a week, fertilize the young plants. Be sure your tomatoes get water once a week.

As the plants get bigger, you'll have to prop them up to keep the growing tomatoes off the ground. When the plants are about 2 feet tall, put a 4- or 5-foot stake in the ground next to each plant. Tie the plant loosely to the stake, using twine. Every few weeks, tie the new branches to the stake. Some gardeners like to put a tomato cage around the plant instead of using a stake. You can buy these cages at most nurseries.

26

TOMATO HORNWORM

You may find tomato hornworms (see the picture on the opposite page) and other caterpillars eating your tomatoes. If they're eating a lot, pick them off and throw them away in a plastic bag. If you have lots of caterpillars, ask an adult to help you spray the plants with BT. This is a pesticide made from bacteria, not chemicals, and it only affects caterpillars. You can buy it at a nursery; it's sold under several brand names.

About ten weeks after transplanting, you should have ripe tomatoes. They'll be a rich golden orange color and will feel slightly soft when you press them. Pick

them carefully by holding the branch of the plant with one hand and the tomato with the other, and pulling gently. Your plants will usually give you tomatoes throughout the summer.

In the fall, listen to the weather reports for your area carefully. If a frost is predicted, be sure to harvest the rest of your tomatoes right away.

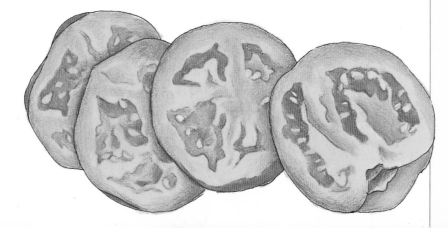

Orange and Red Tomato Swirls

2 large orange tomatoes, washed
2 large red tomatoes, washed
4 tablespoons salad oil
2 tablespoons wine vinegar
6 leaves fresh basil, chopped, or ½ teaspoon dry basil

Cut the tomatoes into thick slices. Find a pretty plate and make a colorful pattern by placing the tomato slices in a ring on the plate. Put down a red slice, then an orange one, and repeat all around the plate, overlapping the slices a little. Put the oil, vinegar, and basil in a small bowl. Whisk with a fork until well mixed, and drizzle over the tomato slices. Serves 4 as a side salad.

Yellow Zucchini

Yellow zucchini squashes look like spears of golden sunlight in the garden. If you want to grow yellow zucchini, send for "Gold Rush" or "Golden Zucchini" seeds from Seeds Blum, Burpee, or Nichols.

If you live in a cold-winter area with a short summer, ask at your local nursery whether you should start your seeds indoors. Otherwise, you can plant outdoors in the spring.

Zucchini plants produce zillions of squashes, so only two plants are needed for most families. To wind up with two plants, make one hill and plant six seeds about 6 inches apart and 1 inch deep. Water the soil well. Then, when the plants come up, water them once a week if it doesn't rain.

When the plants are 3 inches high, pull out all but the two healthiest ones. Fertilize with fish emulsion; then continue to fertilize once a month.

If you live in the eastern United States, you may have problems with squash vine borers (see the picture below), caterpillars that like to eat the middles out of zucchini plant stems. Sometimes they kill part or all of the plant. If part of a plant suddenly looks very wilted, ask an adult to help you take a knife and cut into the stem. If you see what looks like sawdust, carefully poke around for the caterpillar and take it out. Seal it in a plastic bag and throw the bag away. Put the zucchini vine back on the ground and heap dirt over the wounded part, and usually the plant will heal itself.

Your zucchini plants will take about two months from seed planting to produce your first zucchini. Start to harvest the zucchini when they are 4 to 6 inches long. You can cut off the squash right at the stem. Your plants will continue to produce zucchini all summer.

Did you know that zucchini *flowers* are good to eat, too? After your plant has given you lots of zucchini, you can start taking off some of the flowers. Wash, chop, and sprinkle them over salads or soups for an unusual topping.

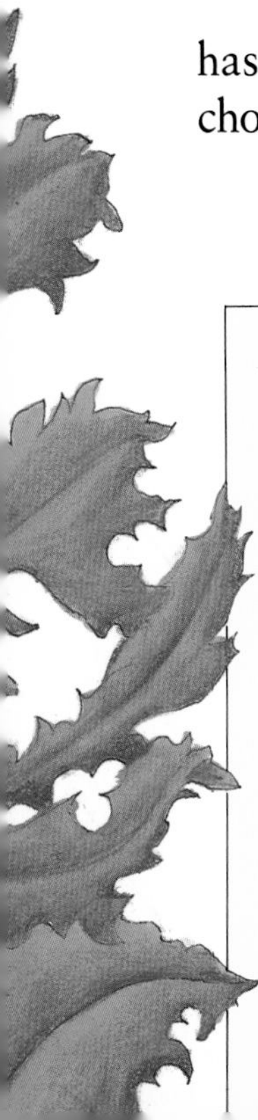

Sunshine Zucchini Muffins

¾ cup flour, sifted
2½ teaspoons baking
    powder
1 tablespoon sugar
1¼ cups yellow cornmeal
¾ teaspoon salt
1 egg

1 cup milk
2 tablespoons salad oil
1 cup grated yellow
    zucchini
1 tablespoon finely
    chopped onion

Preheat the oven to 425°F. Place the flour, baking powder, sugar, cornmeal, and salt in a medium-size bowl. Stir with a spoon. Put egg, milk, oil, zucchini, and onion in a small bowl and mix with a spoon. Then pour the mixture over the dry ingredients and mix lightly.

Grease muffin pan cups with vegetable shortening, and fill the cups half full of batter. Bake for 20 to 25 minutes. Makes 8 to 10 muffins.

SQUASH VINE BORER

Red Chard

Chard, sometimes called Swiss chard, is one vegetable that may not be familiar to you. You may have seen it with the spinach and mustard greens in the produce section of the grocery store. It's a delicious vegetable you can either cook like spinach or cut up raw in salads.

Some chard has white stems and green leaves. But red chard has beautiful ruby-colored stems that can be as brilliant as stained glass. To get this kind of chard, order "Ruby Chard" or "Red Chard" seeds. Most of the seed companies listed on page 9 carry them.

Chard is easy to grow. It will do well in either cool or fairly warm weather, though it sometimes can't stand very hot summer temperatures. You can start planting the seeds outdoors in early spring, as soon as the soil is not frosty or muddy. You can even plant as late as two months before your first fall frost date and still have chard to pick.

To plant your seeds, dig a row 4 feet long. Plant the seeds 1/2 inch deep and 6 inches apart, and

Fancy Red and Green Salad

½ head of lettuce (any kind), washed
2 medium-size tomatoes, washed
2 big red chard leaves (6 inches long), washed

½ cup Homemade Salad Dressing (see page 33) or bottled Italian salad dressing

Take four salad plates and arrange a few lettuce leaves on each plate. Cut each tomato into four wedges and put two wedges on each plate. Cut the stems off the chard and slice each stem lengthwise into two long pieces. Then cut the long pieces into 1-inch sections. You will now have a little pile of chard stems the size of matchsticks.

Tear the chard leaves into 1-inch pieces and sprinkle them over the lettuce. Take a small handful of the red chard stems and sprinkle some on each salad. Drizzle a bit of salad dressing on each. Serves 4.

water them gently. Then water lightly every day until the chard plants come up. Water once a week after that, if it doesn't rain. When the plants are 3 or 4 inches tall, thin some of them out, leaving four healthy plants that are about 1 foot apart. (The tender little plants you pull out can be used in a salad.) Fertilize the remaining plants with fish emulsion every six weeks.

Start harvesting your chard when each plant has six or eight big leaves about 6 to 8 inches long. Pick only two leaves from each plant. In a few weeks, your plants will have grown more leaves, so you can take off two or three more from each of them. You can keep doing this until the weather gets too hot or (if you planted late in the summer) until the weather gets cold and frosty. Ruby Chard can stand light frost and will live all winter in mild-winter areas.

Purple String Beans

Imagine — *purple* string beans! My children used to call these "magic beans" because they're purple when you pick them but turn green as they cook. All the seed companies listed on page 9 carry purple string bean seeds. Look for "Royalty" or "Purple Romano" in your catalogs. Both of these grow on little bushes.

Bean plants take only a short time to grow. Start them in the spring, after the weather turns warm — above 60°F (15°C). (Bean seeds will rot in cool soil.)

Dig one row 10 feet long, or two rows 5 feet long. Your rows should be about 2 feet apart. Plant the seeds 1 inch deep and 2 inches apart, and water the soil well. Each bean plant produces only a few dozen beans, and only for a few weeks. So if you want more beans, you should plant another row or two, in a different part of the garden, a few weeks later.

Once the plants are up and growing, water them once a week if you don't have rain. When the plants are about 3 inches tall, thin some of them out so that the plants you have left are about 4 inches apart. Fertilize when the plants are 6 inches high.

You probably know that red ladybugs with black spots are helpful in the garden because they gobble up aphids and other greedy insects. But the ladybugs' cousins, the Mexican bean beetles (see the picture below), can be a problem. They can

MEXICAN BEAN BEETLE

eat lots of holes in your bean plant leaves. If you see just a few of them, knock them into a bowl filled with soapy water. If you have a lot of them, you can get rid of them with soap spray.

Your string beans will be ready for picking when they are 3 to 4 inches long, about two months after planting. Make sure to pick all the beans over 4 inches long every couple of days. Don't let any of them get big and fat, or the plant will stop producing beans too soon.

Homemade Salad Dressing

5 tablespoons olive oil or salad oil
3 tablespoons wine vinegar
1 teaspoon honey

½ teaspoon (total) chopped fresh herbs such as basil, tarragon, or thyme
Dash of salt and pepper

Pour oil into a small bowl. Slowly dribble vinegar in while whisking with a fork. Add honey, herbs, salt, and pepper, and mix well. Makes ½ cup.

Confetti Bean Salad

1 cup thin carrot rounds
1 cup green bean pieces
1 cup young purple string bean pieces
1 cup water

½ cup Homemade Salad Dressing (see above) or bottled Italian salad dressing

Wash and peel carrots and slice into thin rounds (enough to make 1 cup). Wash the green beans and the purple string beans and cut into 1- or 2-inch lengths (enough to make 1 cup of each color). Set aside the purple bean pieces to add later — don't cook them!

Bring the water to a boil in a medium-size pan. Add the carrots and green beans and cook for 4 to 5 minutes. Drain the water off and put the cooked vegetables into a large bowl. Chill if desired. When you're ready to serve, stir in the salad dressing. Add the raw purple string beans right before serving so that the vinegar in the dressing doesn't turn them green. Serves 4.

33

Multicolored Radishes

When you plant "Easter Egg" radishes, you never know what colors you're going to get! Pink, rose, white, purple, and lavender — all come out of one seed package. Send for "Easter Egg" radish seeds in the early spring or fall. All the seed companies listed on page 9 carry them.

Radishes are fast-growing vegetables. From planting seeds to eating radishes takes only a month, so if you want radishes over many weeks, you have to plant them many times. Because radishes grow best in cool weather, you can start planting them in early spring, as soon as the soil is no longer frosty or muddy. You should not plant them after daily temperatures consistently reach 70°F (21°C) or above. You can also plant them in the fall, up to one month before your average first frost date.

Dig a row 4 feet long and plant the seeds 1/4 inch deep and 1 inch apart. Gently water the soil well, and be sure to keep it moist until the seeds sprout. After that, the plants need watering (by you or by the rain) every four or five days until they're ready to pick.

When your plants are 1 inch tall, thin out some of them, leaving healthy plants that are about 2 inches apart. Fertilize them with fish emulsion.

You might have a problem with tiny flea beetles. These are so small you usually can't even see them; all you see are the little holes they make all over the leaves. Most of the time they don't hurt the radishes; they just make the leaves look bad. If the beetles do get out of control and you can't grow healthy radishes, the next time you plant you might want to cover your seedbed. You can use spun polyester material bought from a nursery.

Another problem you might have is radishes that taste too peppery-hot. This sometimes happens if you forget to water them enough or if the weather gets hot.

You can begin harvesting your radishes when they are 1/2 inch wide. (Their shoulders poke out of the ground as they grow, so you can check their size.) When they get to be more than 1½ inches wide, they get tough and may not taste good.

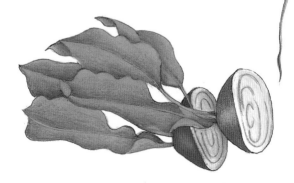

Radish and Carrot Nibbles with Creamy Dip

8 ounces cream cheese or light cream cheese, softened
1 to 3 tablespoons milk
1 tablespoon finely chopped fresh dill
1 teaspoon honey
10 to 12 radishes
4 medium-size carrots

Make the dip by putting the cream cheese, 1 tablespoon milk, dill, and honey in a small mixing bowl and mixing with a fork until smooth. If the dip seems too stiff, you can add more milk (as much as 2 more tablespoons).

Wash the radishes and carrots and drain them. Cut the green tops off the radishes, but leave about 1 inch of stem on each radish to use as a handle when you dip it. Cut the carrots in half crosswise; then cut into long strips. Serves 4 as a snack.

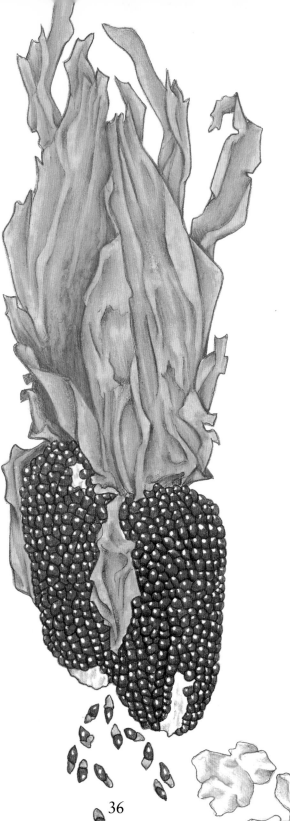

Red Popping Corn

Corn tastes delicious lots of ways — on the cob, ground up into cornmeal to make muffins or cornbread, or popped as a snack. Though we usually think of corn as yellow or white, it can also be red or blue.

"Strawberry" popping corn is especially fun to grow. The plump red ears are pretty in Thanksgiving decorations, and all winter long you can pop the kernels to make white popcorn. You can get "Strawberry" corn seeds from Seeds Blum, Burpee, and Nichols seed companies. This variety will give you kid-size plants about 4 feet tall.

Corn plants take almost four months to make corn, so if you live in a cold-winter area with a short summer, ask at your local nursery if you should start your seeds indoors. Otherwise, you can begin planting outdoors in the spring.

In order for your corn plants to produce fat cobs filled with kernels, the wind must blow pollen from the tassels onto the silks (the long "hairs" around the young ears of corn). This is much more likely to happen if your plants grow in a group, rather than in a long, skinny row, so be sure to plant your corn in hills.

To get twenty to forty cobs, make four hills. Plant ten seeds in each, about 6 inches apart and 1½ inches deep. Then gently water the soil well. Birds sometimes dig up the seeds or young plants, so put bird netting (available from a nursery) over the hills. Once your plants are up and growing, water them thoroughly once a week if you don't get rain.

When your plants are 3 inches tall, pull out all but five healthy plants in each hill. (Each plant will give you one or two cobs.) Try to leave one in the middle and four around the outside. Take the bird netting off when the plants are 4 inches tall. At this point, fertilize with fish emulsion. Fertilize the plants at least two more times — once when the plants are 8 to 10 inches tall, and again when the silks first appear.

Sometimes caterpillars (see the picture below) get into the tops of corn cobs and nibble off a few kernels. This won't really hurt your corn. But if it bothers you, try carefully opening the corn tops and picking out the caterpillars. Seal them in a plastic bag and throw the bag away. Be sure to push the husks back over the corn cobs so they won't dry out.

In about three to four months, your corn plants will start to turn yellow; then they'll turn brown. Once they're brown, take the cobs off and peel back the husks to help them dry. Put them in a warm, dry place indoors for about a month.

You can use the cobs for decoration, if you wish. To save the kernels afterward, remove them by pushing them with your thumbnail or twisting the cob really hard in your hand. (If you need help, ask an adult.)

Store the kernels in a tightly closed jar.

CATERPILLAR

Quick-and-Easy
Microwave Popcorn

½ cup popping corn   1 tablespoon butter

Place the popping corn in a brown paper lunch bag. Close the bag loosely by folding the top over twice. Put it in the microwave and turn the power to high. Cook for about 3 minutes. The time may vary, depending on your microwave; you might have to experiment a bit. When the corn kernels stop popping, open the bag. (Be careful — hot steam will come out the top!) Put the butter in a small bowl, cover it with a plate, and put it in the microwave. Turn it on low power (about 3) for a few minutes until the butter is melted. Pour the melted butter over the popcorn. Close the bag again and shake to coat the popcorn evenly. Makes 4 cups of popcorn.

Yellow Watermelon

There's nothing better than a slice of juicy yellow watermelon on a hot summer day.

That's right — yellow! Yellow watermelons make a wonderful addition to your garden, but remember that they need a lot of room to grow and a lot of hot weather to be sweet. You can write to Burpee, Harris, or Park seed companies to get "Yellow Baby Hybrid" seeds. This variety is kid-size — each fruit is about as big as a cantaloupe. The yellow flesh tastes the same as red watermelon, but "Yellow Baby Hybrid" has fewer seeds.

It will take about ten weeks for your seeds to sprout and produce melons. If you live in a cold-winter area with a short summer, ask someone at your local nursery whether you should start your watermelon plants indoors. Otherwise, you can plant them right outside in the spring, after the soil and weather are nice and warm and there is no chance of frost.

Watermelons are grown in hills. To get two healthy plants, plant six seeds in a hill, about 6 inches apart and 1 inch deep. Each watermelon plant generally gives two or three fruits. If your family really loves watermelon and you have space, you'll want to plant two hills of them. Watermelon plants can spread to 4 or 5 feet wide, so give each hill lots of room.

When the plants are 3 inches high, pull all but the two healthiest ones out of each hill. Fertilize with fish emulsion. In a month, fertilize again. If you don't get much rain, water your watermelon plants once a week.

After about ten weeks, check to see if your watermelons are ripe. This is a little tricky, and you may not always get it right. Here are a few hints: Usually the shiny skin turns dull and gets tough enough so that it can't be slit with your fingernail. Some people thump on a watermelon with their knuckles; they say it goes "plink" when it's ripe, but "thud" when it's not. Or you can roll the melon over a bit and peek underneath at its belly. If it's quite yellow, that's a good sign that it's ripe.

Rainbow Fruit Salad

| 1 yellow watermelon | 12 strawberries |
| ½ cup blueberries | Fresh mint sprigs |

Chill the watermelon in the refrigerator for several hours. Ask an adult to cut the chilled watermelon in half and show you how to make watermelon balls using a melon scoop. (Don't worry if they aren't perfect — you can always eat your bloopers!) Scoop out the melon flesh with the melon scoop and place the little melon balls in a large bowl. (Don't scoop the greenish flesh right next to the skin, because it might be bitter.)

Wash the blueberries and strawberries and drain them in a colander. Remove the green caps from the strawberries and slice the strawberries. Add the blueberries and strawberries to the melon balls and mix carefully. Divide among four sherbets or other glass dishes. Tuck a couple of mint leaves alongside the fruit in each dish. Serves 4.

Index